It's All About Physics

Nicolas Brasch

Australia • Brazil • Japan • Korea • Mexico • Singapore • Spain • United Kingdom • United States

It's All About Physics

Fast Forward
Emerald Level 25

Text: Nicolas Brasch
Editor: Johanna Rohan
Design: Stella Vassiliou and Ami Sharpe
Series design: James Lowe
Production controller: Seona Galbally
Photo research: Gillian Cardinal
Audio recordings: Juliet Hill, Picture Start
Spoken by: Matthew King and Abbe Holmes

Acknowledgements
The author and publisher would like to acknowledge permission to reproduce material from the following sources:
Front cover: iStockphoto/Don Bayley (light globe)
Back cover: Photolibrary (atomic structure)
iStockphoto: p 4 bottom/Andrew Johnson, p 5 top/Christoph Ermel, 5 middle/Don Bayley; Jupiterimages: p 7 bottom; Photolibrary, pp 4 top & middle, 5 main, 6-7, 7 bottom, 8. All other photographs by Lindsay Edwards, except p 22-23 by Ami Sharpe.

ISBN 978 0 17 012724 0
ISBN 978 0 17 012717 2 (set)

Cengage Learning Australia
Level 7, 80 Dorcas Street
South Melbourne, Victoria Australia 3205
Phone: 1300 790 853

Cengage Learning New Zealand
Unit 4B Rosedale Office Park
331 Rosedale Road, Albany, North Shore NZ 0632
Phone: 0508 635 766

For learning solutions, visit cengage.com.au

Printed in Australia by Ligare Pty Ltd
7 8 9 10 11 12 13 21 20 19 18 17

THE UNIVERSITY OF MELBOURNE

Evaluated in independent research by staff from the Department of Language, Literacy and Arts Education at the University of Melbourne.

Nicolas Brasch

Contents

INTRODUCING PHYSICS

Physics is the study of matter and energy. Matter is anything that occupies space.

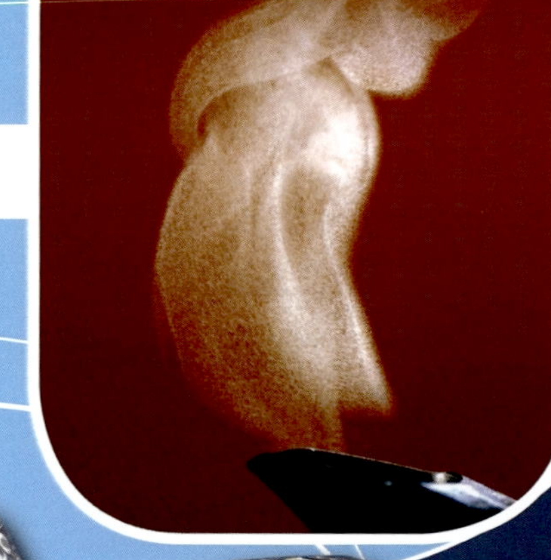

Gases are matter.

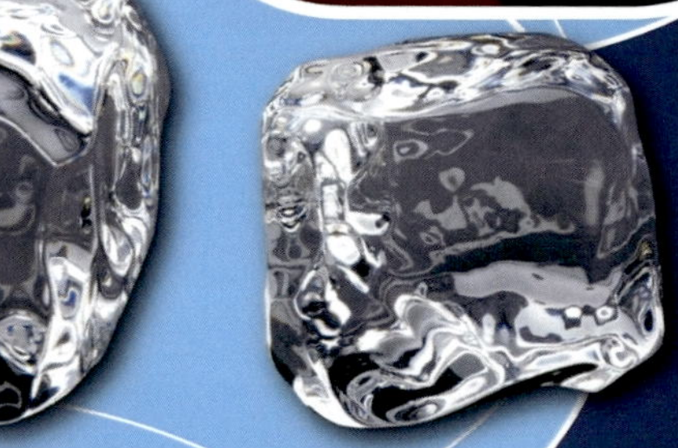

Solids are matter.

Liquids are matter.

Physics is the study of everything that exists – the way it acts, the way it **reacts** and the way it moves.

We use physics to explain how things work. Physics helps to explain what the universe is made of, how it began and how everything in the universe acts.

Physics can help to explain how turning on a light switch causes a globe to light up. It can also help to explain how heavy aeroplanes are able to move through the air without falling. Scientists who study physics are called physicists.

Chapter 2

BRANCHES OF PHYSICS

There are many branches of physics. Physics is also a major part of many other important areas of science, including electronics, engineering and medicine.

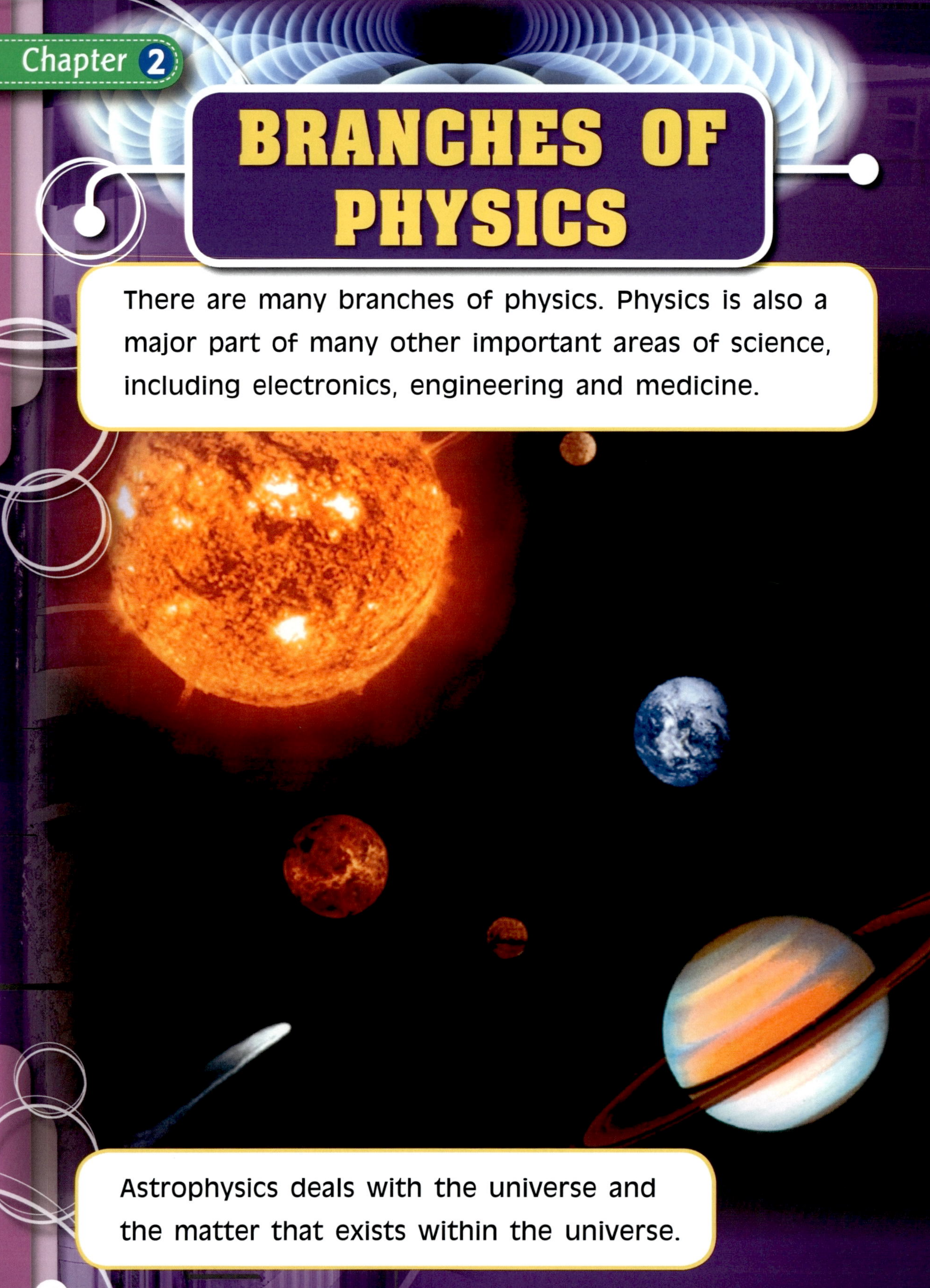

Astrophysics deals with the universe and the matter that exists within the universe.

Biophysics looks at the way physics works within living organisms. A biophysicist may study the way muscles and nerves move and are controlled.

Fluid and plasma physics looks at the movement and properties of fluids and plasmas. Plasmas are gases.

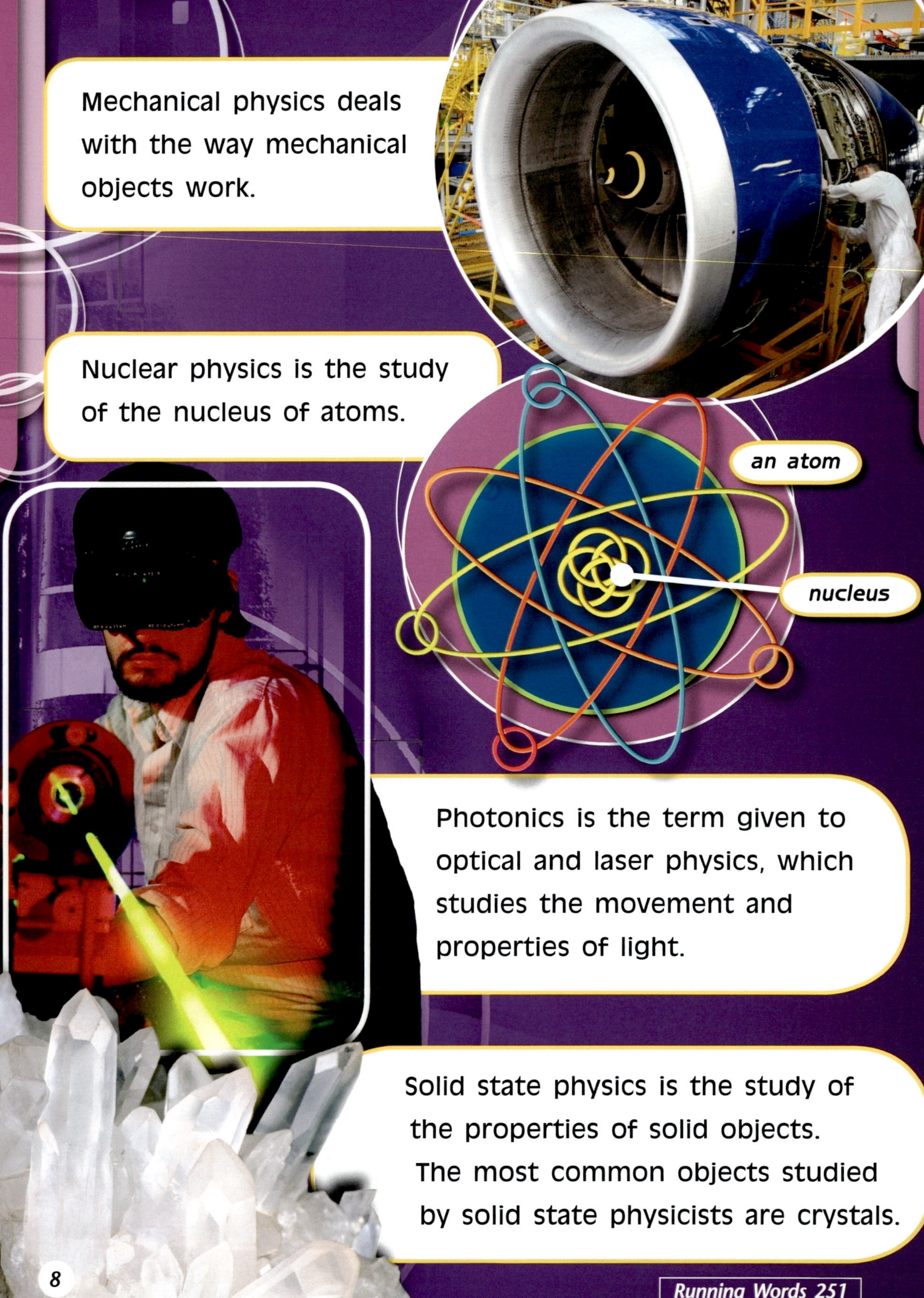

Mechanical physics deals with the way mechanical objects work.

Nuclear physics is the study of the nucleus of atoms.

Photonics is the term given to optical and laser physics, which studies the movement and properties of light.

Solid state physics is the study of the properties of solid objects. The most common objects studied by solid state physicists are crystals.

Running Words 251

Chapter 3

BIOPHYSICS EXPERIMENT

Aim

To test a person's reflexes.

Materials

- a metre-long ruler
- a chair

Procedure

1. One person should stand on the chair, holding the ruler vertically in front of them.
2. Another person should stand near the ruler and put their fingers close to the 50 centimetre mark, without touching the ruler.

3. Without warning, the person on the chair should let go of the ruler.
4. The other person should try to catch the ruler.
5. Make a note of the position where the ruler was caught.
6. Swap places and repeat the experiment.

Observation

The person's reaction time is measured by the position of their hand after they have caught the ruler.

Conclusion

The person who grabs the ruler closest to the 50 centimetre mark has the fastest reaction time.

This experiment can also be used to see how people's reactions are different at various times of the day. Do the experiment in the morning and again before bed. See if there is any difference.

Chapter 4

MECHANICAL PHYSICS EXPERIMENT

Aim

To prove that a lever makes it easier to pull one object out of another object.

Materials

- a piece of wood
- a nail
- a hammer with a claw end

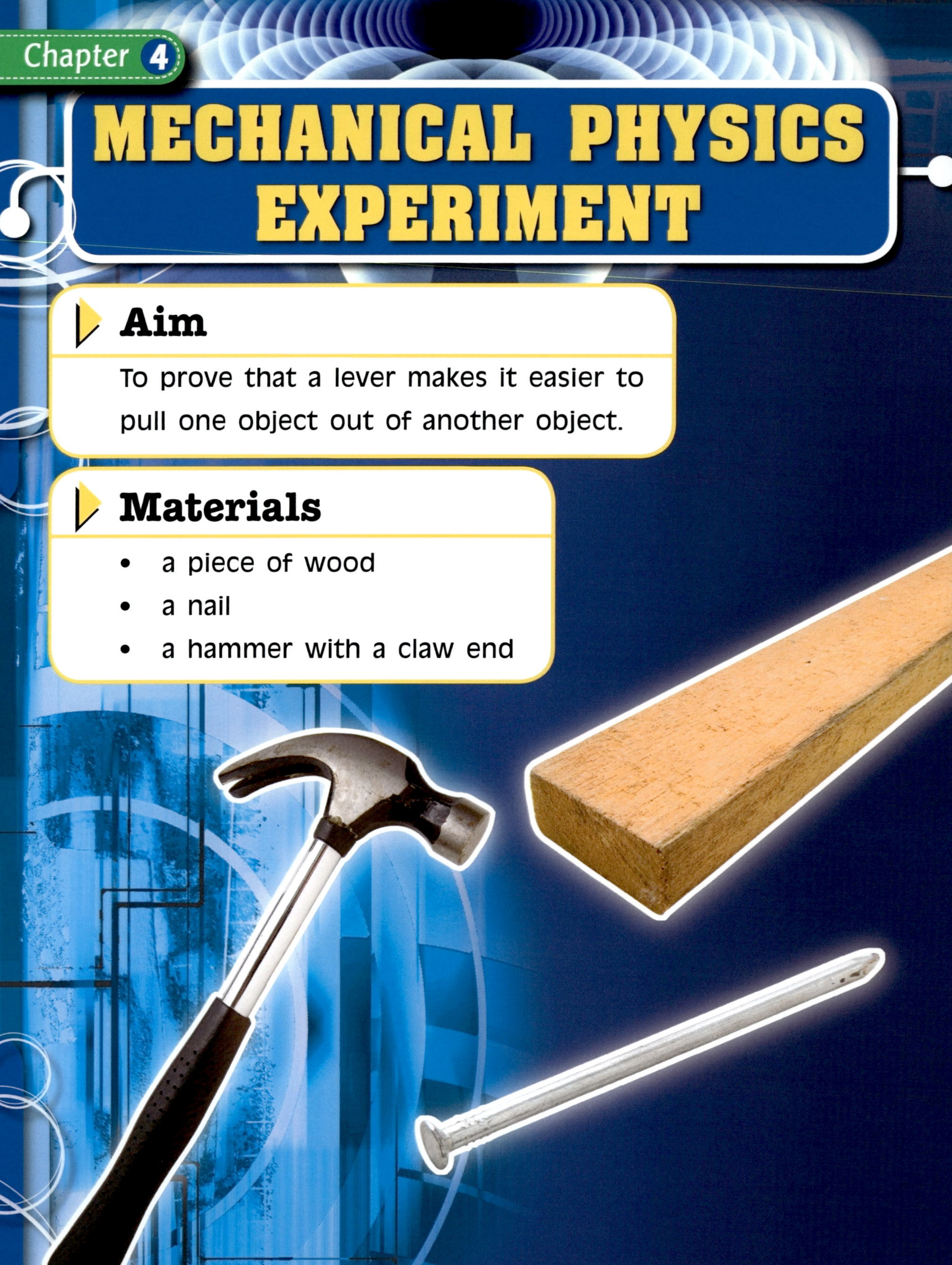

Procedure

1. Hammer the nail into the piece of wood.
2. Try to pull the nail out of the wood with your hand. If you have hammered it in far enough, it shouldn't move.

3. Now, use the claw end of the hammer. Place it so that the nail is sticking up through the middle of the claw.
4. Place the curved part of the end of the hammer against the wood.
5. Rotate the curved end so that it slowly loosens the nail. Continue until the nail comes out.

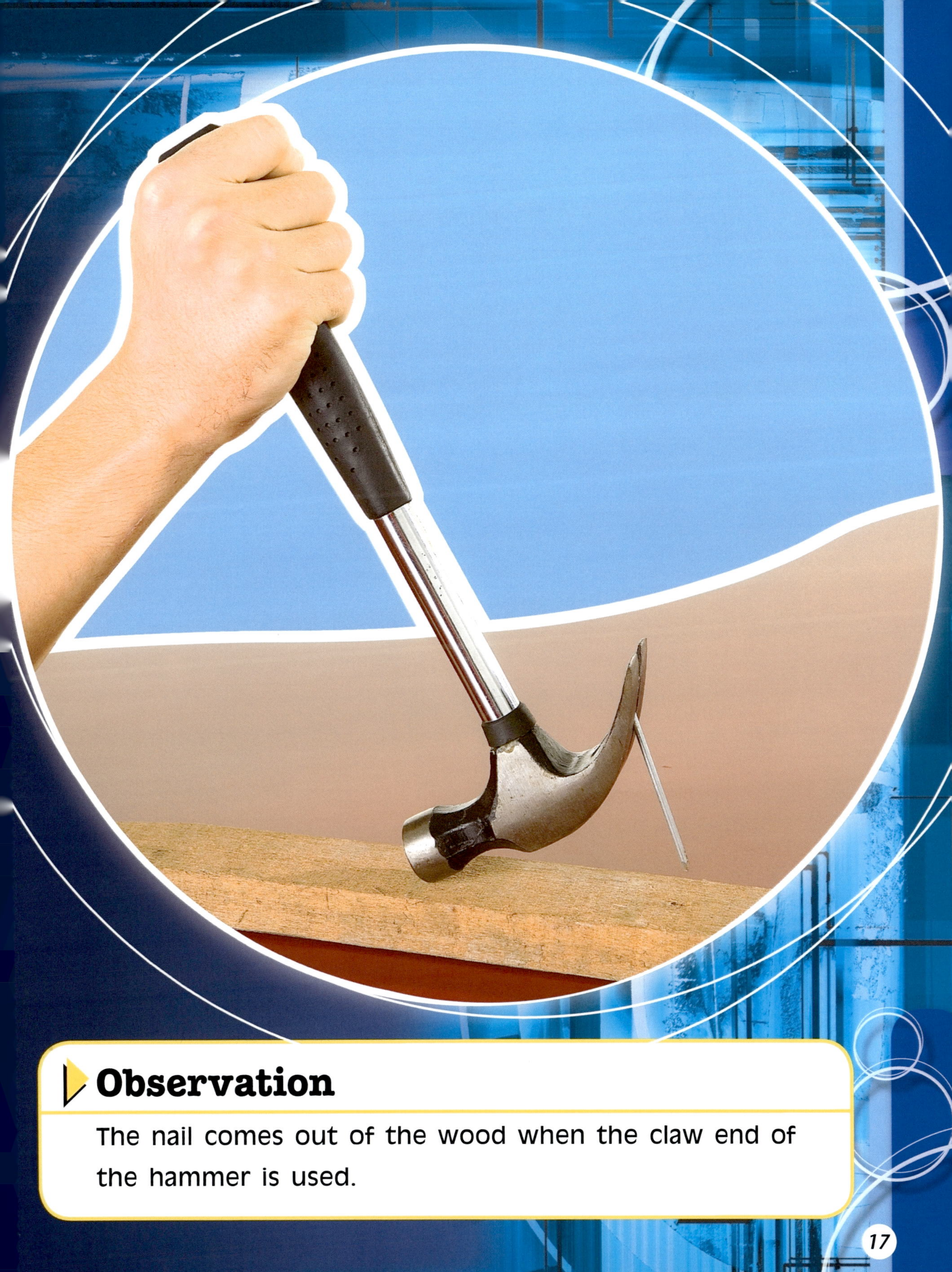

Observation

The nail comes out of the wood when the claw end of the hammer is used.

Conclusion

It requires less effort to pull a nail from a piece of wood when a lever is used than when you try to pull it out using **brute force**.

A lever is an arm that **pivots** against a **fulcrum**. Levers can be used to move very heavy objects with little effort.

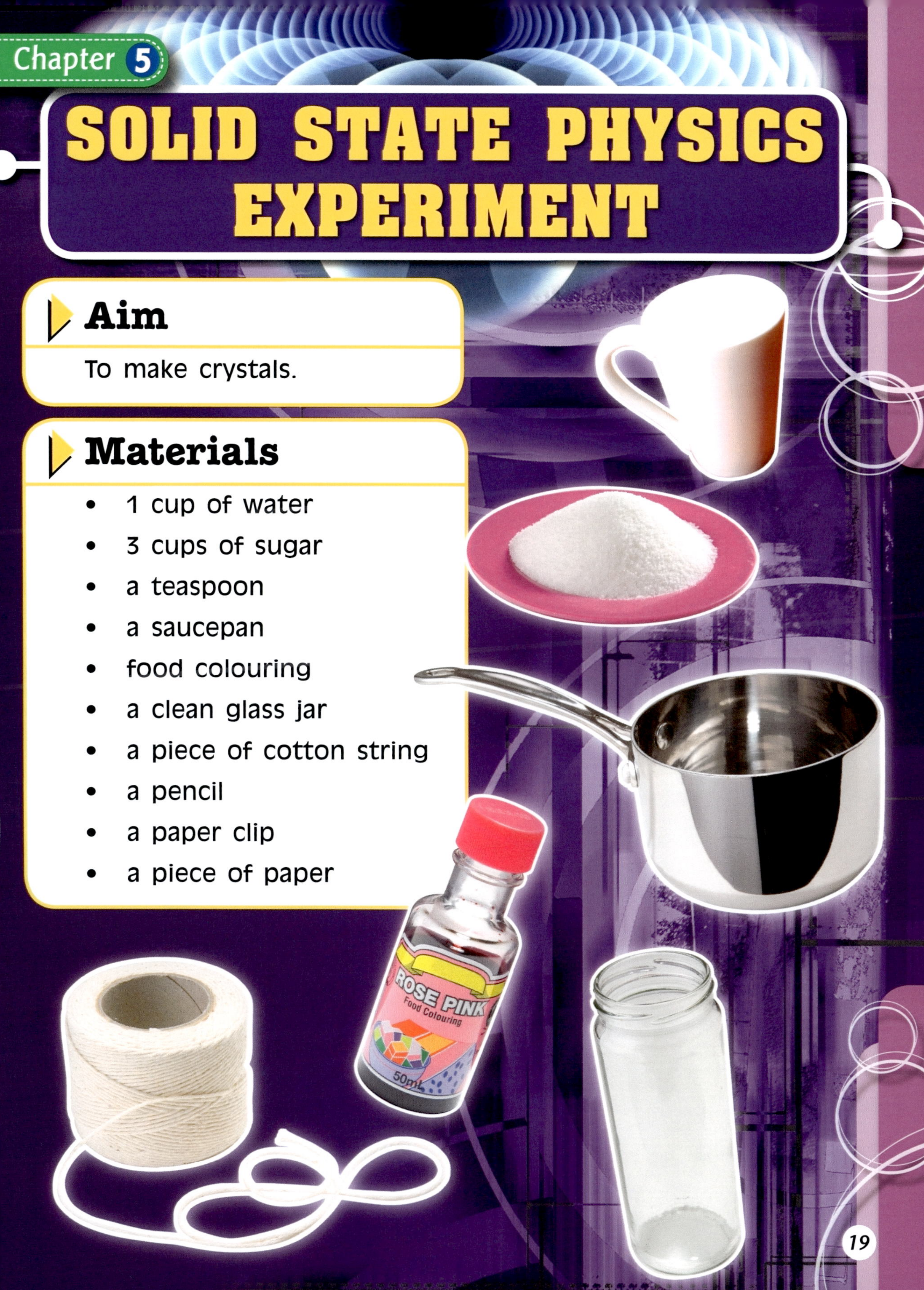

Chapter 5

SOLID STATE PHYSICS EXPERIMENT

Aim

To make crystals.

Materials

- 1 cup of water
- 3 cups of sugar
- a teaspoon
- a saucepan
- food colouring
- a clean glass jar
- a piece of cotton string
- a pencil
- a paper clip
- a piece of paper

Procedure

1. Boil the water in the saucepan.
2. Slowly stir the sugar into the water, one teaspoon at a time. Don't rush this step.
3. Continue step two until the sugar is no longer dissolving. It will start to collect on the bottom of the saucepan.
4. Put a few drops of food colouring into the water.
5. Pour the solution into the glass jar.

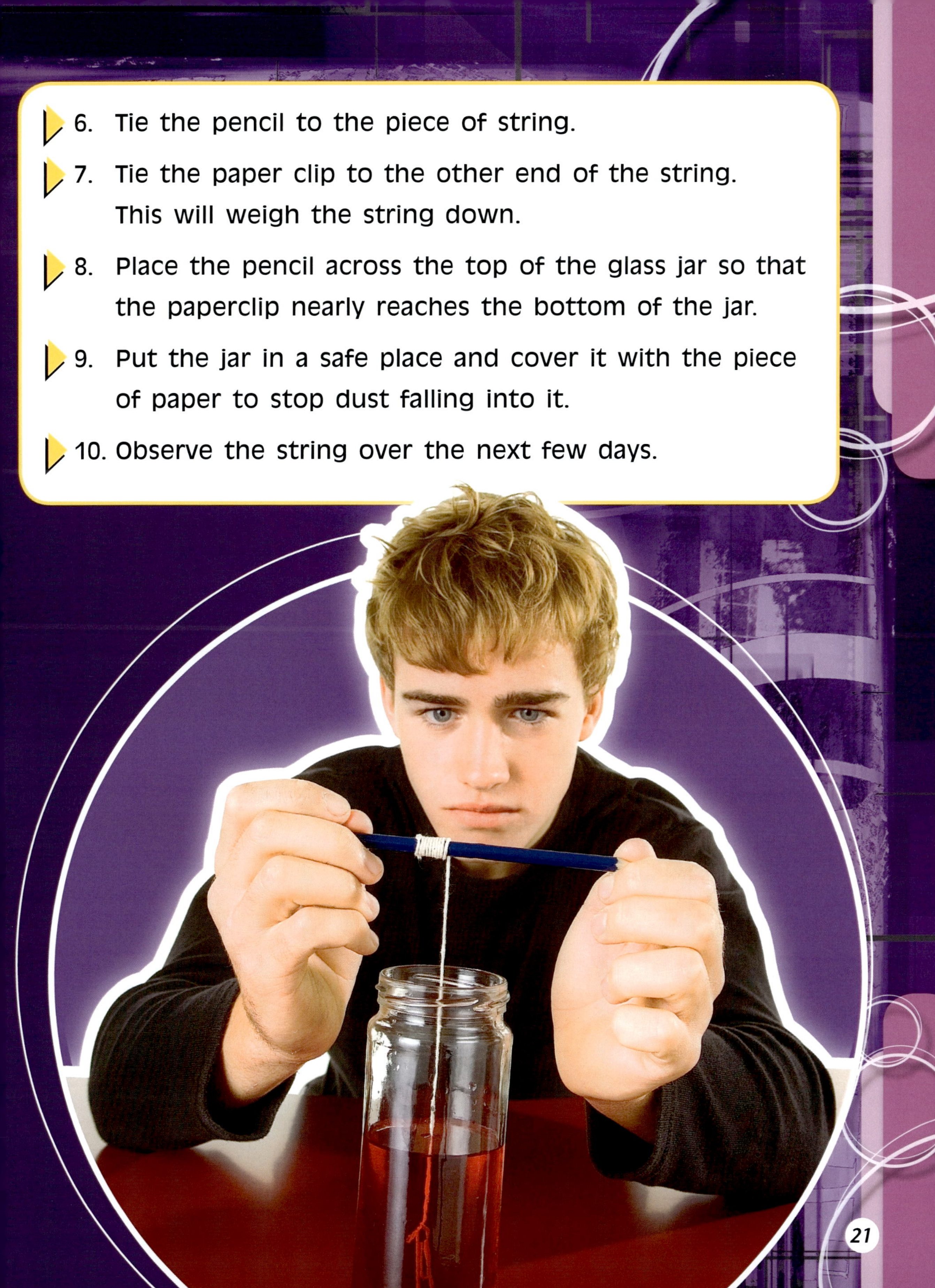

6. Tie the pencil to the piece of string.
7. Tie the paper clip to the other end of the string. This will weigh the string down.
8. Place the pencil across the top of the glass jar so that the paperclip nearly reaches the bottom of the jar.
9. Put the jar in a safe place and cover it with the piece of paper to stop dust falling into it.
10. Observe the string over the next few days.

Observation

Crystals start to form on the string. This could take a few days to a week. They will continue to form. Because they are made from sugar and have food colouring to make them easier to see, they can be eaten.

Conclusion

While making crystals this way doesn't take long, it takes millions of years for the Earth's **minerals**, like salt or carbon, to make a crystal. Salt can also be used in this experiment instead of sugar.

Glossary

brute force physical effort

fulcrum a point of support on which a lever turns in raising or moving something

minerals natural compounds formed through geological processes

pivots turns

reacts responds

Index